The Day AIDS Began March 14, 1927

The Making of a Hypothesis

George E. Parris

Dedication

"Hypotheses are not statements of truth, but instruments to be used in the ascertainment of truth. Their value does not depend upon ultimate verification, but is to be measured by their effect upon scientific research."

...C. Stuart Gager, University of Missouri 1909

(Translator's notes on *Intracellular Pangenesis*, 1910)

Preface

My interest in HIV grew out of my interest in how parasites prevent cells they invade from committing suicide. I had been studying apoptosis and in particular the functions of the protein Bcl-2. As soon as I read that the cells that are infected with HIV-1M kill similar cells (bystanders)[1] that are not infected with HIV, I immediately suspected that the virus was fortifying infected cells against apoptosis and probably making their cytokine response more potent. The result was that the infected cells were inadvertently killing off *uninfected cells* in an immune system war of attrition. Before I read this fact, I just assumed that the virus was killing the host cells that it invaded.

In any event, the idea that humans were the only primates in the Congo Basin without a resident immune virus seemed anthropocentric. And of course, why

---

[1] Garg H et al. 2012. HIV-1 Induced Bystander Apoptosis. *Viruses.* 4(11): 3020–3043.

would primates harbor these types of viruses? Somewhere along the line it occurred to me that malaria was the greatest threat to life in the Congo Basin and it was clear that many adverse mutations were being accumulated in the human population for the sole reason that the mutations reduced the pre-pubescent mortality (which even today runs 30-50%). Extreme threats require extreme defenses.

An examination of the life-cycle of malaria pointed to the liver stage as the most important bottleneck; defeat malaria there and you can beat malaria. But, how to do that? The malaria parasites do exactly the same thing as other intracellular parasites. When they take over liver cells, they suppress the apoptosis system and turn the liver cells into nurseries for the next stage of the parasite (which invaded the oxygen/hemoglobin-carrying packets of the blood…erythrocytes). Thus, it seems likely (and I have hypothesized)[2] that T-cells fortified with HIV

---

[2] Parris GE. 2004. Hypothesis links emergence of chloroquine-resistant malaria and other intracellular pathogens and suggests a new strategy for treatment of diseases caused by intracellular parasites. *Med. Hypothesis*. 62(3):354-7.

Parris GE. 2007. HIV-infection of CD4+ T-cells enhances their effectiveness against hepatocytes infected with malaria parasites. *Med Hypotheses*. 68(5):1187-8.

infections are better at attacking these malaria-infected liver cells.

The idea that HIV would be a harmless passenger virus had been proposed by Peter Duesberg, who has been savaged by his critics.[3] My departure from the view of Duesberg was that it seemed reasonable that resistance to a drug could develop in a non-target "organism" (e.g., a virus).[4] And my first though was chloroquine[5], which was

---

[3] Duesberg is a well-known academic who stubbornly resisted the notion that every unexplained infection observed in the early days of the AIDS epidemic should be attributed to HIV. He associated these infections with a lifestyle of drugs and sex; and discouraged the government actions that have been taken to mitigate the spread of HIV. Thus, he became and has remained an anathema to the homosexual community and international health organizations.

[4] Parris GE. 2007. AIDS: caused by development of resistance to drugs in a non-target intracellular parasite. *Med Hypotheses*. 68(1):151-7.

[5] Parris GE. 2006. Update on hypothesis linking chloroquine-resistant malaria to disease-causing HIV. *Med Hypotheses*. 67(3):670-1.

introduced into the Congo Basin circa 1955[6] in an attempt to eradicate malaria by the World Health Organization. At that time (1955-60), virtually everyone in the Congo Basin had started receiving large doses of chloroquine.

But work by a series of researchers[7], was pointing towards an origin for HIV-1M around 1930 or earlier. It did not take long to figure out that chloroquine was not

---

[6] Parris GE. 2006. The timing is right: Evolution of AIDS-causing HIV strains is consistent with history of chloroquine use. *Med Hypotheses.* 67(5):1258-9.

[7] Korber B et al. 2000. Timing the ancestor of the HIV-1 pandemic strains. *Science.* 288(5472):1789-96.

Hillis DM. 2000. AIDS. Origins of HIV. *Science.* 288(5472):1757-9.

Salemi M et al. 2001. Dating the common ancestor of SIVcpz and HIV-1 group M and the origin of HIV-1 subtypes using a new method to uncover clock-like molecular evolution. *FASEB J.* 15(2):276-8.

Worobey M et al. 2008. Direct evidence of extensive diversity of HIV-1 in Kinshasa by 1960. *Nature.* 455(7213):661-4.

invented until 1934 and was not even used in the Congo until the mid-1940s. Thus, chloroquine could not have selected for mutations before 1930.

A large number of acrimonious public debates and accusations had arisen in the early 2000s regarding the origin of HIV-1M. Edward Hooper, a journalist, had published a book (*The River*) with an interesting hypothesis that HIV had its origins in development and testing of the polio vaccine in the Congo Basin in the 1950s. Initially, the scientists had welcomed him into their discussions and they tested his hypothesis. But when it was obvious that he was wrong, he refused to give up the limelight and started throwing around accusations of scientific fraud and coverup. At the time, the predominant (though not the only) seriously considered theory of HIV-1M origin among academics was the so called "cut hunter" theory.

The idea of the cut hunter was that that a hunter had cut himself while butchering an SIV-infected chimpanzee (*Pan t t*) somewhere in extreme south eastern Cameroon (on the Sangha River) and this act *alone* was adequate to cause minor mutations in the SIV to produce HIV-1M. The biggest problems with this theory are that these zoonoses (transfer of SIV from an animal to humans) have happened many times in many settings; yet HIV-1M is the only infection causing anything that looks like a pandemic. Nonetheless, the cut hunter hypothesis has no western villains and was supported by some of the leaders in the academic circle studying HIV.

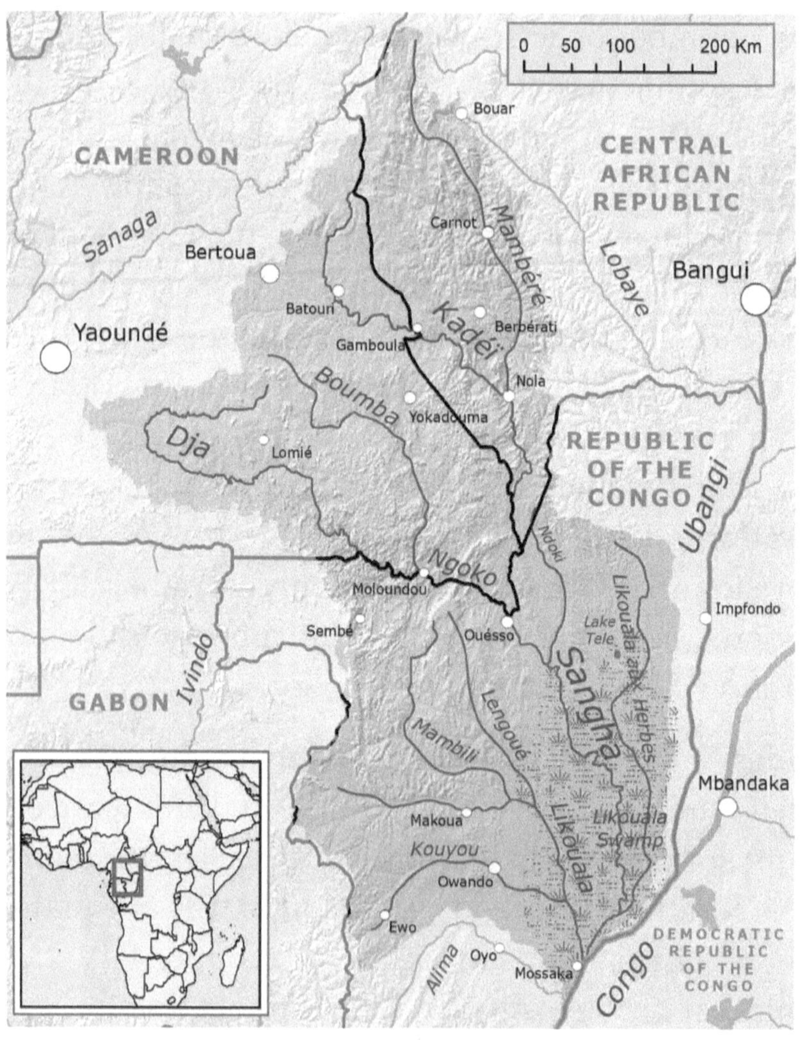

Source:
https://en.wikipedia.org/wiki/Sangha_River#/media/File:Sanghabasinmap.png

It seemed to me that there was an effort in the academic community to "close the gap" between the last SIV infection and the first HIV-1M infection (i.e., the most recent common ancestor of HIV and SIV) to support the cut hunter hypothesis. But the data were not working out that way. The evidence shows that (as I would expect) humans have been infected with HIV long before the arrival of HIV-1M (i.e., the most recent common ancestor of HIV-1M and the closest SIV is sometime in the 1800s or earlier).

The other major hypothesis from the recognized academicians was the serial transfer hypothesis involving unsterilized immunization needles.[8] The problem with that hypothesis has to do with the timing of the events and the expansion of the HIV-1M-infected population. It turns out that there were massive screening programs for sleeping sickness in the Congo and Cameroon in the 1920s and there is evidence that some viral infections were spread during this activity (through reuse of needles). But HIV-1M was not spread.

Indeed, one of the points I will focus on here is that there seems to be a 10-15-year lag between the establishment of

---

[8] Parris GE. 2007. Comments on the serial-passage hypothesis of HIV evolution. *Med Hypotheses.* 2007;68(4):914.

the first few strains of HIV-1M (subtypes A, J, and all the other subtypes) and expansion for the *number of people* infected. The phenomenon is called the starburst:

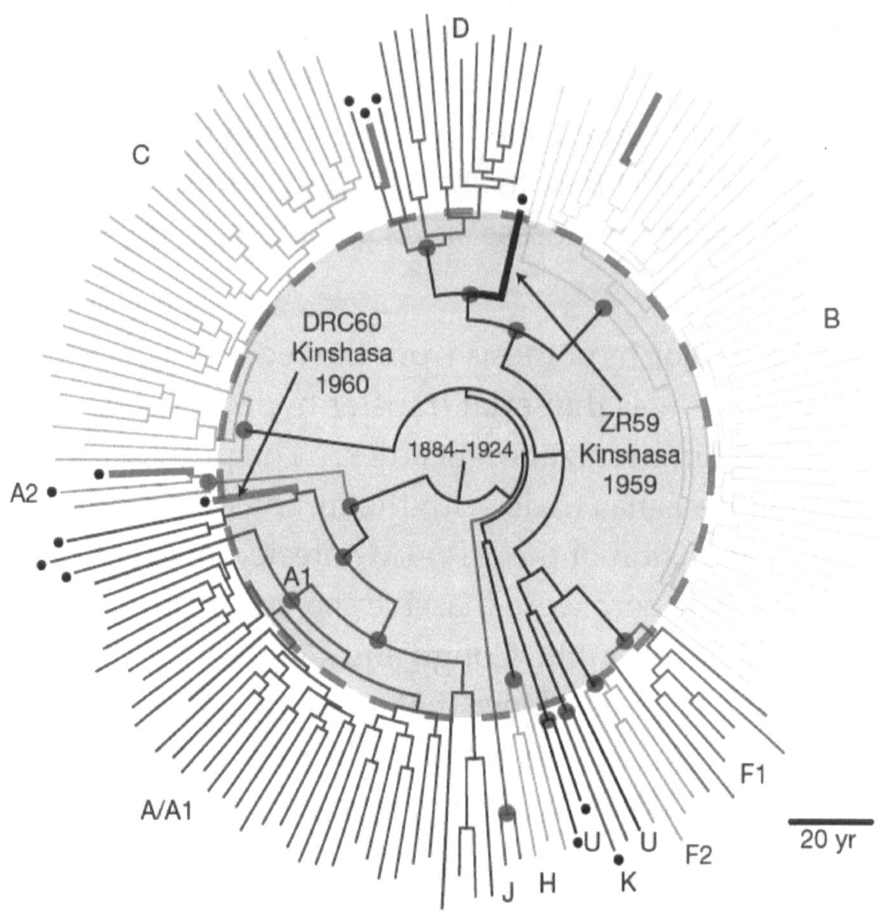

Source: Worobey M et al. 2008. Direct evidence of extensive diversity of HIV-1 in Kinshasa by 1960. *Nature*. 455(7213):661-4.

You will notice that the center of this figure is dated "1884-1924." And as late as the 1950s, only a hand full of strains are known (i.e., the principle subtypes of HIV-1M). But there is a major explosion of HIV types in the late 1950s (concurrent with the introduction of chloroquine, which may be totally coincidental). But the fact is that through the 1920s, to the 1940s, there were very few strains of HIV-1M. This tends to rule out serial transfer (which was clearly happening in the 1920s and 1930s with the massive screenings for sleeping sickness) as a mode of initiation of HIV-1M. Indeed, it tends to rule out the presence of HIV-1M in the adult population (who were clearly engaging in sexual activities constantly through these time periods). More about that later.

Thus, I continued to look for a scenario that would explain the data at hand (2006-2007) and started digging back into the medical literature. The history of the Congo Basin especially in the period 1870-1910 is brutal and a shameful chapter in European colonialism. The international slave trade was banished, but Africans were virtually enslaved in their own country. I recommend reading the book *Heart of Darkness* (1899) by Joseph Conrad to get a sense of the times. The Congo River (unlike the Mississippi, the Amazon, the Nile or the Yangtze) does not flow smoothly into the Atlantic Ocean. The bulk of the Congo River sits on a high plateau (the

Congo Basin) and the tremendous rain fall passes slowly and collects at the lowest point in the western ridge where it all spills over through a series of steep rapids to the ocean. Because the lower Congo is not navigable, the Belgians built a narrow-gauge railway (using conscripted African labor) for about 200 miles from Matadi on the coast to Leopoldville (Kinshasa) at a lake originally called Stanley Pool (now Lake Kinshasa).

Tropical diseases were killing off Europeans at a steady rate (as they also kill off Africans) and the Belgians built a clinic at Kinshasa, as early as about 1905. By 1920, this clinic was being expanded and was the place where Louise Pearce came (1921) to conduct trials on sleeping sickness. (more about this in a moment).

I began poking around and found that the Congo was one of the places that tropical medicines of all sorts were typically tested; and the clinic at Leopoldville (Kinshasa) was the focus of this activity. Looking for drugs that might have been tested in the 1920s, I found my way to the *Bulletin De La Societe De Pathologie Exotique* and found that its indices were actually online on the internet. In the index, I found an interesting lead, a drug (plasmoquine; a.k.a. pamaquine) was tested at the clinic in early 1927. I found an e-mail address and sent a request to the library in Belgium. To my surprise and delight, I had a digital photocopy of the paper on my computer within a week.

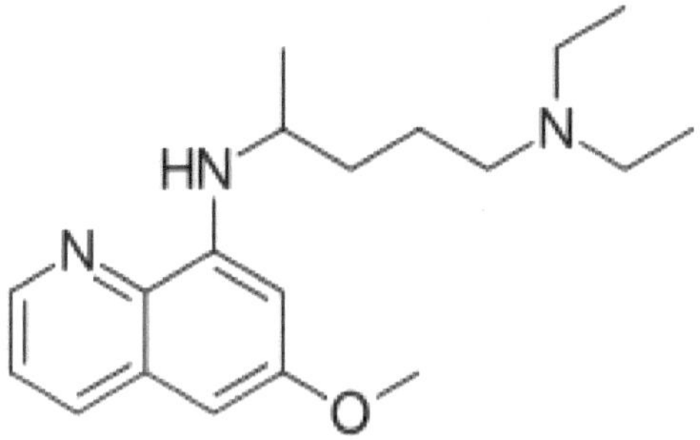

Pamaquine (a.k.a. plasmoquine)

Author: Fvasconcellos; Source: Wikimedia Commons

The drug had been synthesized in 1924 and tested against malaria in birds in 1926. The paper described a small test on native African children in early 1927. The drug was tested in India by the British Army on a larger scale in 1929 and it was still in use during the early stages of WWII. It is rather toxic and was replaced by primaquine in 1946. The advantage of pamaquine was that it did attack the liver form of malaria and could prevent relapses. But the side effects were bad enough to keep the US Army looking for a drug during the War and they discovered that chloroquine (first synthesized in 1934, but overlooked by the Germans) was superior in most respects.

I stared at this paper for a long time. It was in French and not easy for me to decipher. It looked to be from the right place and the right time and I really had not though about the issue before, but all the patients to whom it was administered were young girls (under 10 years old). As an isolated data point, it was a long, long leap to associate the test of pamaquine with the origin HIV-1M.

Then two things happened: In my review of the work of Louise Pearce in Kinshasa (1920), I found a letter in the archives of the National Library of Medicine on the Internet.[9] In this letter to her colleague Michael Heidelberger regarding her 1920 clinical trial of tryparsamide (A63), she mentions that there is no sleeping sickness locally, but most of her patients are being brought in by steamboat from 500 miles upriver.

> *17 July 1920*
>
> …
>
> *"The director* [Dr. Par F. Van den Branden] *has charge of the negro hospital in which there are about 70 advanced cases of sleeping sickness undergoing the routine atoxyl & emetic treatment – 0.5 g atoxyl on Monday – 0.1 emetic on Thursday. I was given 3 relapses to begin on – a most unfortunate type of case – but the next day, an Early untreated patient arrived. 2.0 grams of* [tryparsamide] *cleared the cervical lymph glands of trypanosomes within 21 hours and I am now waiting to see when a relapse takes place.*

---

9 https://profiles.nlm.nih.gov/ps/retrieve/ResourceMetadata/DHBBNQ

I hope to get more similar cases from ***the nearest epidemic center—about 500 miles away!*** *"*

She goes into more detail in her published manuscript.[10]

The object of the mission sent to the Belgian Congo was the obtaining of concrete facts and observations as outlined above. All of the work was done at the government laboratory and Hôpital de la Reine in Léopoldville. The helpful cooperation of the Belgian and Colonial Governments made it possible to accomplish a considerable amount of work in a comparatively short time and it is a pleasure to express here our appreciation of the aid so cordially extended to us, first in Brussels and then in the Congo. We also wish to acknowledge the assistance given us by Professor Rodhain, Chief of the Colonial Medical Service, and by Dr. van den Branden, Director, and Dr. van Hoof, Assistant Director of the Laboratory in Léopoldville.

SOURCE AND TYPE OF PATIENTS.

The observations which form the basis of this report were made on 77 native cases of trypanosomiasis caused by *Tr. gambiense* which were treated with tryparsamide in Léopoldville, Belgian Congo. The

[10] Pearce L. 1921. Studies on the treatment of human trypanosomiasis with Tryparsamide (the sodium salt of N-phenylglycineamide-p-arsonic acid). *Journal of Experimental Medicine*. xxxiv(6) supplement 1: 1-104.

civil service of the Belgian Congo has established throughout the colony, medical stations and posts under the charge of government physicians and traveling sanitary agents, in addition to which a considerable amount of medical work is done by various religious missions. The native population is thus examined for trypanosomiasis at more or less regular intervals, the routine procedure consisting of the palpation of the cervical lymph glands and the microscopic examination of lymph gland juice from those individuals having one or more palpable cervical glands. At the present time, there is no widespread epidemic of a severe character in the immediate vicinity of Léopoldville but there are many endemic foci of the disease in this district, while a few days' travel brings one into active epidemic areas. The patients treated with tryparsamide were entirely typical of the routine run of trypanosomiasis cases met with in Léopoldville. The population of the nearby native villages comes *en masse* to the laboratory every 3 months for examination and several cases were obtained in this way. Moreover, Léopoldville is the terminus of one of the large river transportation companies and a medical passport issued after examination for trypanosomiasis, is required of any native traveling from one part of the colony to another. The entire crew of each boat is examined the day before sailing and this source yielded a number of cases from distant parts of the colony. Finally, several advanced cases under treatment in the native Hôpital de la Reine and lazaret in Léopoldville were transferred to us for treatment with tryparsamide. All of the patients studied thus fell into three general classes: First, those who were sent to the lazaret, which is in reality a native village under nominal native police control on the outskirts of Léopoldville, and who came to the laboratory when sent for; second, patients in the Hôpital de la Reine, over whom a closer supervision could be maintained; and third, ambulatory cases, corresponding to dispensary patients, who continued to live in their own villages and who came to the laboratory at certain fixed times.

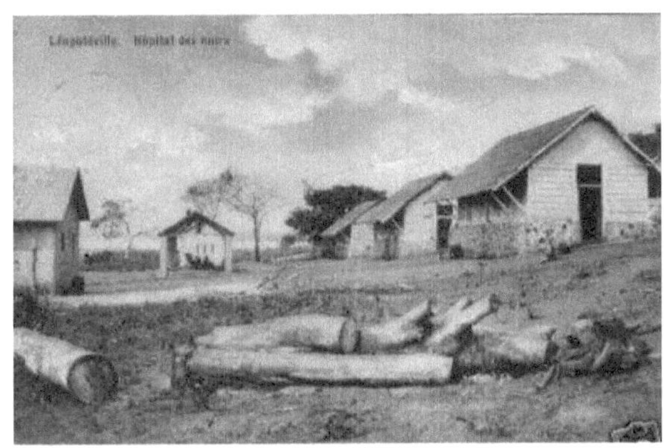

Hospital de la Rive

Lazaret

Photo Source: Kinshasa Then and Now
http://kinshasa101.rssing.com/chan49414006/all_p1.html

Leopoldville.
Congo Belge.
17 July, 1920

Dear Dr. Heidelberger –

It was very good of you to send me a book for my journey – and I certainly appreciate your thoughtfulness. As perhaps you know – my steamer packages were not put on board the Kroonland through somebody's stupid mistake – but they were sent on another boat & I got them here. Books are more than acceptable in this part of the world – We get mail only once a month – & the local paper contains nothing but advertisements.

I have had a very strenuous & exhausting journey which I'll tell you about on my return – but at last I am here & actually at work. At times, I thought that Leopoldville must be on another planet. The laboratory building is in the process of construction – but the ground floor of 3 rooms is practically finished. There is no electricity at present - but the permanent equipment is fairly good. The director has charge of the negro hospital in which there are about 70 advanced cases of sleeping sickness undergoing the routine atoxyl & emetique treatment - 0.5 g atoxyl on Monday - 0.1 g emetique on Thursday. I was given 3 relapses to begin on - a most unfortunate type of case - but the next day, an early untreated patient arrived. 2.0 grams of A63 cleared the cervical lymph glands of trypanosomes within 21 hours & I am now waiting to see when a relapse takes place. I hope to get ~~some~~ more similar cases from the nearest epidemic center - about 500 miles away! With best wishes - & renewed thanks for the ...

I am - Sincerely yrs.
Louise Pearce

Nouvelle médication du Paludisme par la Plasmoquine,

Par F. Van den Branden et Mlle E. Henry.

La plasmoquine est un sel d'alcoyl-amino-6-méthoxy-quinoléine obtenu par voie synthétique. Il ne s'agit donc pas d'un produit dérivé de la quinine.

Le médicament se présente sous forme d'une poudre insipide, finement granuleuse, de coloration jaune clair, assez facilement soluble dans l'alcool, soluble dans l'eau dans la proportion de 0,03 o/o, à 20°. L'acide chlorhydrique du suc gastrique transforme facilement la plasmoquine en chlorhydrate.

La plasmoquine nous a été délivrée en comprimés contenant 0 g. 02 du produit, et en dragées contenant 0 g. 01 de plasmoquine et 0 g. 125 de quinine ; cette dernière composition est dénommée plasmoquine composée.

Le médicament est presque absolument dépourvu de saveur et par suite moins désagréable à prendre que la quinine, qualité fortement appréciable en thérapeutique infantile. De plus, elle ne provoque aucun des symptômes désagréables consécutifs à l'emploi de la quinine, tels que bourdonnements d'oreilles, vertiges, etc.

Aux doses thérapeutiques, la tolérance vis-à-vis de la plasmoquine est bonne ; elle produit quelquefois de la tachycardie qui disparaît par simple diminution des doses ou en mettant un intervalle de repos de quelques jours au cours du traitement.

Posologie. — Dans la fièvre quarte ou dans la fièvre tierce bénigne, on donne à l'adulte 0 g. 02 de plasmoquine 3 fois par jour pendant 5 jours, puis 3 jours de repos, puis 4 jours de plasmoquine. Le traitement doit être poursuivi durant 4 à 6 semaines en observant ces intervalles.

Dans la fièvre tierce tropicale avec gamètes dans le sang, on a recours à l'association de la plasmoquine avec la quinine ; plasmoquine composée. On donne, à l'adulte 2 dragées 3 fois par jour, autant que possible sans interruption pendant 1 mois, et la moitié de cette dose, soit 1 dragée 3 fois par jour, durant le mois suivant.

Les enfants tolèrent des doses très élevées du produit ;

au-dessus de 10 ans, on peut essayer de leur faire prendre la même dose que celle administrée à l'adulte. Pour les enfants, en dessous de 10 ans, les doses doivent être diminuées de moitié ou du tiers suivant l'âge.

Nous avons employé la plasmoquine dans la fièvre tierce tropicale et dans la fièvre quarte. Les cas de fièvre tierce bénigne étant rares à Léopoldville, il nous a été impossible d'essayer l'action du nouveau produit sur le *Plasmodium vivax*. La posologie indiquée précédemment n'a pas pu être strictement observée au cours des observations, ci-après relatées.

I. — Action de la Plasmoquine sur les schizontes de tierce tropicale (« Plasmodium falciparum »)

Observation n° 1. — Ntongo Stéphanie, enfant âgée de 5 ans pesant 17 kg. 600. Atteinte de trypanosomiase chronique et traitée au tryponarsil. Présente dans le sang des schizontes de tierce tropicale.

Reçoit du 3-2-1927 au 12-3-1927, 3 dragées de plasmoquine composée par jour, excepté le samedi et le dimanche où elle ne reçoit que 2 dragées.

Le 14-3-27, nous faisons un examen du sang en goutte épaisse et nous trouvons encore des parasites de tierce tropicale. A cette date, le poids est de 17 kg.

Observation n° 2. — Jeanne, enfant âgée de 3 ans du poids de 12 kg. 300. Schizontes de tierce tropicale dans le sang.

Reçoit du 3-2-1927 au 12-3-1927, 3 dragées de plasmoquine composée par jour, excepté le samedi et le dimanche, où nous donnons 2 dragées.

Le 14-3-27, nous trouvons à l'examen du sang en goutte épaisse des schizontes de tierce tropicale. Le poids passe à 11 kg. 300.

Observation n° 3. — Bula, Albertine, enfant âgée de 8 ans environ pesant 16 kg. 400. Son sang renferme des schizontes de tierce tropicale et des gamètes.

Reçoit du 4-3-1927 au 14-3-1927, tous les jours, 3 dragées de plasmoquine composée.

L'examen du sang en goutte épaisse pratiqué le 14-3-1927 révèle la présence de schizontes de tierce : nous ne trouvons plus de gamètes. A cette date, le poids est de 16 kg. 300.

Du 15-3-1927 au 22-3-1927, l'enfant reçoit encore tous les jours 3 dragées de plasmoquine composée.

Le 22-3-1927, un nouvel examen du sang démontre la présence de schizontes de tierce tropicale et la disparition des gamètes.

La plasmoquine ne paraît pas avoir d'action sur les *schizontes* de tierce tropicale.

I was still looking at the paper by Dr. Branden (1927) who was Dr. Pearce's host in Leopoldville. And suddenly it struck me, sleeping sickness was endemic in south western Cameroon and the Sangha River area. The first child in the pamaquine study was named "Ntongo Stephanie" (5 years old in 1927) and she was cured of sleeping sickness (trypanosomiasis) with tryparsamide before she began the treatment with pamaquine for malaria.

Fig. 3. — Laboratoire de Léopoldville. — Salle des cours pratiques.

Bacteriological Laboratory and Leopoldville

A quick search of the name "Ntongo" revealed it to be a common name in Cameroon with place names on the Sangha River (Ntongo-Atani). The child could easily have arrived in Leopoldville (Kinshasa) by steamboat just as Dr. Pearce's patients had found their way to the same clinic.

And on 14-3-1927 (March 14, 1927) she was the first of three young girls (5, 3 and 8 years old) who had their blood drawn to test for the malaria parasite.

This is as much evidence for a hypothesis as anyone else has presented and I managed to get it published through the kind cooperation of Bruce Charlton editor of *Medical Hypothesis*.[11] However, I am somewhat disappointed that (to my knowledge) neither of these papers has ever been

---

[11] Parris GE. 2007. Mechanism and history of evolution of symbiotic HIV strains into lethal pandemic strains: the key event may have been a 1927 trial of pamaquine in Leopoldville (Kinshasa), Congo. *Med Hypotheses.* 69(4):838-48.

Parris GE. 2007. How did the ancestral HIV-1 group M retrovirus get to Leopoldville from southeastern Cameroon? Med Hypotheses. 69(5):1098-101.

cited in papers where they are highly relevant, illuminating and consistent with reported findings.

Dr. Charlton was very tolerant of my numerous papers over those years and he deserves credit for having an open mind. I mention this in particular because shortly thereafter, he published a letter from Peter Duesberg, that was actually a comment on a paper that had appeared in a different journal. I suspect that Carlton did not know that the paper had been submitted to, and rejected by, the other journal and I'm sure that he was not prepared for the absurd complaints that suddenly appears about him and *Medical Hypotheses*. He had stepped into the political whirlwind that followed Peter Duesberg. The AIDS lobby called for his removal (which they got) and they called to have *Medical Hypothesis* removed from the NIH list of journals that were routinely abstracted. Indeed, they wanted all the abstracts removed on the grounds that there were not peer-reviewed.[12] Fortunately, that did not happen.

12 This is where I like to point out that Charles Darwin's *Origin of the Species* was a self-published and un-peer-reviewed document. I share these ideas at my own risk and will be happy to defend my reasoning or change it. *Do not tie your ego to a "fact," tie it to the scientific method.*

The latter-day Analyses of the Origins of HIV-1M

I sense that the interest in the origin of HIV-1M has died down over the last decade. The acrimony is also fading into the past. Edward Hooper still has his website and is still of the opinion that the proof of his hypothesis is out there somewhere. But as best I can tell it has been over three years since anything material has been posted there.

In the meantime, I have stood by waiting for someone to publish work that prove me to be definitively wrong. After all, I predicted the date, place and who was in the room when HIV-1M got its start. To my pleasure, I have not had to bury all those citations in my resume, *yet*. But no one seems interested in citing me either. I suspect that the criticism that *Medical Hypothesis* has received (and the expectation that the people whom various authors did cite might well be on the peer review) motivated the authors' choices of citations.

I am prompted to write this essay because in spite of being generally ignored, recent research seems to be consistent with the hypothesis I published in 2007.

The First Important Paper

The first paper that seems to support this hypothesis was published in 2014.[13] This paper included a long list of well respected "HIV-1M origin" academics centered on Philippe Lemey's group (Belgium) who have published a number of important papers.

If you look at the population dynamics of HIV-1M,

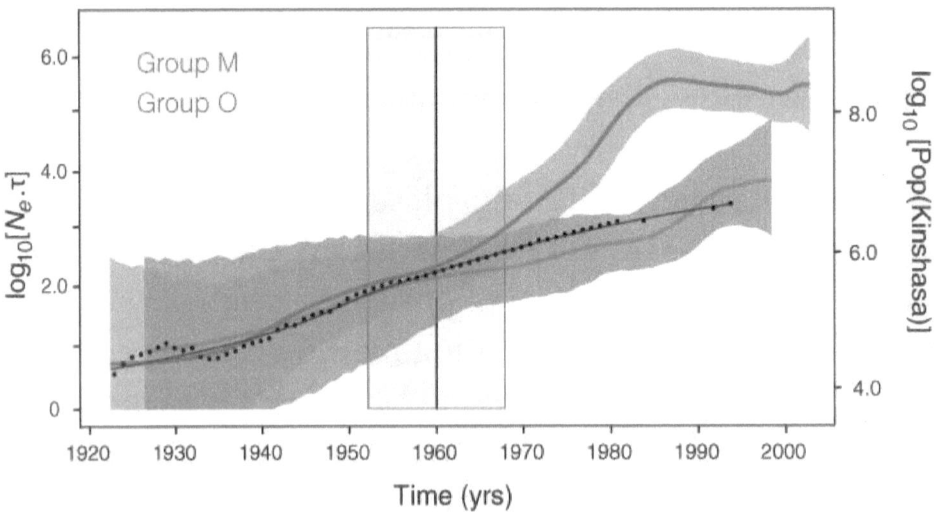

---

[13] Faria NR, Rambaut A, Suchard MA, Baele G, Bedford T, Ward MJ, Tatem AJ, Sousa JD, Arinaminpathy N, Pépin J, Posada D, Peeters M, Pybus OG, Lemey P. 2014. HIV epidemiology. The early spread and epidemic ignition of HIV-1 in human populations. *Science*. 346(6205):56-61.

notice that in 1930, the number of infected people appears to be about $10^{0.7}$ (which is a total of 5 people); it does not reach $10^{1.0}$ (which is 10 people) until nearly 1940. This can fairly certainly be identified as some small group of people who shared some common experience in the late 1920s. And it is clear that something out of the ordinary happened in the late 1950s.

The authors focused on the spread of the infection over time. They confirmed what has been deduced by others: that the epicenter must be Kinshasa (where the only medical facility of note was the Belgian clinic in the 1920s). They conclude that:

> *"The cross-species transmission of SIV to humans predates the group M common ancestor and probably occurred in southeast Cameroon, where the chimpanzees with SIVcpz strains most similar to group M have been identified. After localized transmission, presumably resulting from the hunting of primates, the virus probably traveled via ferry along the Sangha River system to Kinshasa. During the period of German colonization of Cameroon (1884–1916), fluvial connections between southern Cameroon and Kinshasa were frequent due to the exploitation of rubber and ivory."*

I have removed their citations, but note that for their authority for the Sangha River they cite Sharp PM, Hahn BH. *Nature*. 2008;455:605–606 and some of their own authors de Sousa JD, Alvarez C, Vandamme AM, Müller V. *Viruses*. 2012;4:1950–1983 (rather than the earlier paper that I published in 2007, which actually included the information regarding Dr. Pearce, which is fact, not speculation).

Nonetheless, Faria et al. provide another very relevant piece of information:

> *"…our results revealed that the earliest introductions of HIV-1 to Brazzaville occurred by 1937 (95% BCI: 1920–1953). We note that these estimates pertain to viral lineages that survived to be sampled in each location; thus, HIV-1 may have been introduced earlier (e.g., to Brazzaville) but without successful onward transmission. Historical transportation data from the DRC during 1900–1960 suggests that viral lineages in migrant populations living in or around Kinshasa would have had many opportunities for introduction to DRC regions connected to other population centers in central Africa."*

They also note that Brazzaville is only 6 km from Kinshasa. What? It took over 10 years for the virus to move 6 km, really? Well this would not make sense if the virus was carried by an adult, but if the virus was carried only by children who in 1937 were merely 13 to 18 years old, it would be reasonable. Of course, none of my work is cited.

The Second Important Paper

The second paper worth mentioning was published early in 2019.[14] The authors used a very clever technique to look for likely recombination in the "original" HIV-1M genome. What they found was clear evidence of recombination of an old strain of the virus that dated from about 1913 (95% CI 1906-1918), and a younger strain dating from 1929 (95% CI 1924-1933). Notice that the two confident intervals do not overlap and are much narrower than most others that have been reported. Notice also that the younger segment (representing about 5000 nts at the 5′ end of the genome) is evolving more slowly and has an uncertainty that includes the date that I proposed in 2007 (i.e., early 1927). The older segment primarily representing the *env* gene has been evolving faster at least since the recombination. The authors note that the substitution rate for the extreme 3′ end of the

---

[14] Olabode AS, Avino M, Ng GT, Abu-Sardanah F, Dick DW, Poon AFY. 2019. Evidence for a recombinant origin of HIV-1 Group M from genomic variation. *Virus Evol.* Jan 22;5(1) vey039.

genome (the *nef* gene) is similar to the rate in the newer recombinant segment.

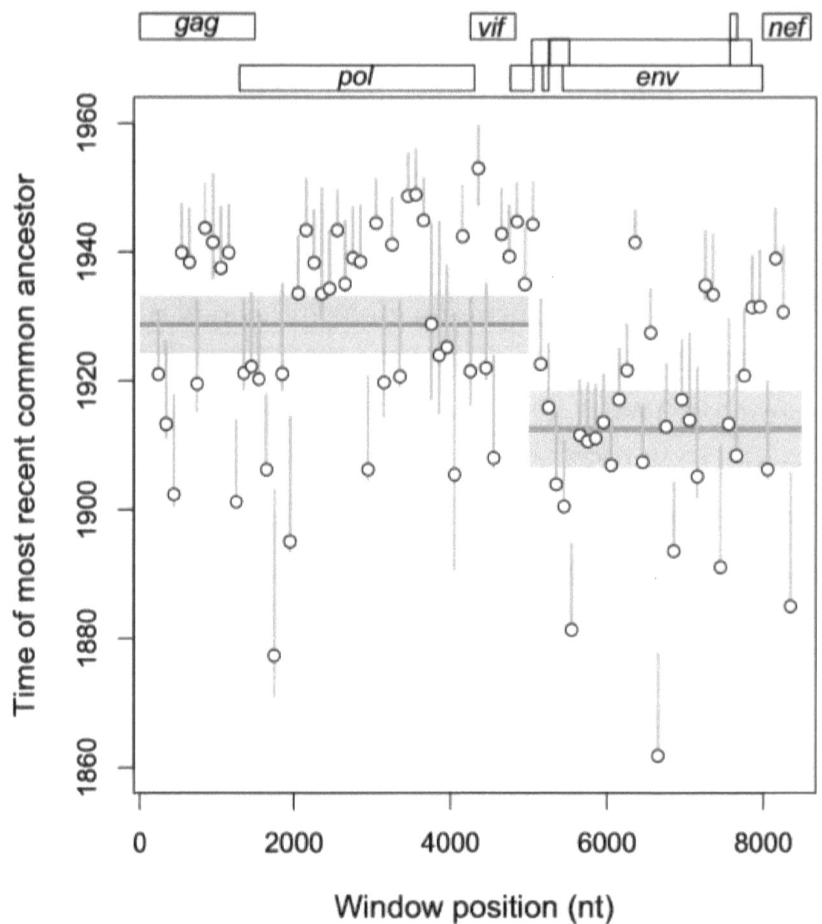

Source: Olabode AS, Avino M, Ng GT, Abu-Sardanah F, Dick DW, Poon AFY. 2019. Evidence for a recombinant origin of HIV-1 Group M from genomic variation. Virus Evol. Jan 22;5(1) vey039.

The authors convincingly ruled out purifying selection as an explanation for the differences in the ages of the segments. Overall, these results seem to confirm a recombination event in the late 1920s as the root of all HIV-1M viruses. It gives the most precise date published so far and is within two years of the event that could explain the conversion of a relatively harmless HIV virus into a lethal pandemic.

What is the Basis for the Change in Virulence?

While a substantial amount of work has been done on trying to identify the origin of HIV-1M, little progress has been articulated towards understanding why this virus is lethal or how it got that way. Everyone agrees that there was some sort of change in the HIV genome in the late 1920s, but what caused it and what does that tell us?

If you go back to the "cut hunter" hypothesis, the agent of change was simply putting the SIV virus into humans. But this idea is destroyed by the facts that these zoonoses (i) are not rare, (ii) not lethal, and (iii) usually seem to burn themselves out. The zoonosis hypothesis also is inconsistent with the fact that HIV existed in humans long before HIV-1M evolved. Similarly, it is not clear why serial transfer would have led to a more dangerous virus. And it is inconsistent with the timing or location of evolution of HIV-1M.

So, we need a mechanism to get HIV to Leopoldville (Kinshasa) in the late 1920s; and we need a good reason for the virus to mutate in a way that it had never mutated before. And, this apparently happened in people who did not have sex or travel far until the late 1930s. Data have been gathered but they have not been systematically articulated, until now.

The Two-Step
Pamaquine-Chloroquine Selection Hypothesis

Start with this hypothesis:

(i) Humans in the Sangha River area have carried various strains of HIV for many years. These infections generally came from animals by various mechanism (not unlike the "cut hunter" hypothesis). Indeed, these events are common and the infections generally burn themselves out.

I suspect HIV provides some benefit against malaria; i.e., HIV-infections arm CD4 T-cells with higher levels of FasL, which may induce apoptosis in malaria-infected schizonts in the liver.[15] CD8 T-cells use TNF-α to attack liver schizonts.[16] There is evidence that increased immune response is very effective in eliminating malaria parasites in the liver. [17] And induction of apoptosis of infected hepatocytes seems to be the mode of action of artemisinin. [18]

(ii) One of the children that was included in the pamaquine trial in 1927 at the Kinshasa clinic was

---

15 Malyshkina A et al. 2017. Fas Ligand-mediated cytotoxicity of CD4+ T cells during chronic retrovirus infection. *Sci Rep.* 7(1):7785.

16 Butler NS et al. 2010. Differential effector pathways regulate memory CD8 T cell immunity against *Plasmodium berghei* versus *P. yoelii* sporozoites. *J Immunol.* 184(5):2528-38.

17 Emran TB et al. 2018. Baculovirus-Induced Fast-Acting Innate Immunity Kills Liver-Stage *Plasmodium*. *J Immunol.* 201(8):2441-2451.

18 Mo P et al. 2014. Apoptotic effects of antimalarial artemisinin on the second generation merozoites of *Eimeria tenella* and parasitized host cells. *Vet Parasitol.* 206(3-4):297-303.

Jiao J et. al. 2018. Artemisinin and *Artemisia annua* leaves alleviate *Eimeria tenella* infection by facilitating apoptosis of host cells and suppressing inflammatory response. *Vet Parasitol.* 254:172-177.

apparently from this area and was also infected with trypanosomes and malaria parasites. She was cured of the trypanosomes before entering the malaria/pamaquine trial.

(iii) During this trial, one of the unrecognized effects of pamaquine was to kill off certain classes of immune T-cells. For example, pamaquine causes hemolytic anemia in people with glucose-6-phosphate dehydrogenase deficiency (G6PDD).[19] And this may have been associated with changes in the T-cells, which are essential for immunization against malaria.[20]

---

[19] Zhang P et al. 2013. An in vivo drug screening model using glucose-6-phosphate dehydrogenase deficient mice to predict the hemolytic toxicity of 8-aminoquinolines. *Am J Trop Med Hyg*. 88(6):1138-45.

Berman J et al. 2018. Tafenoquine and primaquine do not exhibit clinical neurologic signs associated with central nervous system lesions in the same manner as earlier 8-aminoquinolines. *Malar J*. 6;17(1):407.

[20] Azizi G et al. 2016. Autoimmunity in primary T-cell immunodeficiencies. *Expert Rev Clin Immunol*. 12(9):989-1006.

Primaquine

In the process, the drug selected for the most robust (apoptosis-resistant) immune cells and thus selected for the most powerful anti-apoptotic strain of HIV.[21]

Smirnova SJ et al. 2016. Expansion of CD8+ cells in autoimmune hemolytic anemia. *Autoimmunity.* 49(3):147-54.

Allison AC and Eugui EM. 2016. The role of cell-mediated immune responses in resistance to malaria, with special reference to oxidant stress. *Annu Rev Immunol.* 1983;1:361-92.

[21] Recent reports concerning the lncRNA SAF emphasize the importance to fortifying the HIV-infected cells against apoptosis.

Boliar S et al. 2019. Inhibition of the lncRNA SAF drives activation of apoptotic effector caspases in HIV-1-infected human macrophages. *Proc Natl Acad Sci U S A.* 116(15):7431-7438.

Villamizar O et al. 2016. Fas-antisense long noncoding RNA is differentially expressed during maturation of human erythrocytes and confers resistance to Fas-mediated cell death. *Blood Cells Mol Dis.* 58:57-66.

Although Van den Branden did examine changes in the relative numbers of major types of blood cells in later experiments, there is no real evidence for this supposition in his work (1927).

(iv) One of the preferred features of reduced inclination to apoptosis is suppression of transcription of pro-viruses. Thus, the selected (1927) HIV pro-virus was rarely transcribed and evolved slowly. Nonetheless, it appears that it did manage to recombine with an older faster-evolving virus (dating from ~1913, see above).

(v) The recombined virus selected by pamaquine HIV_{1927} was spread among a very small group (3 individuals) of young (3 to 8 years) African girls during a trial of pamaquine in Kinshasa. They remained in Kinshasa and did not begin having sex until their teens (approximately 10 years). The virus, thus, only spread to their sex partners and children until approximately 1945 when

---

Villamizar O et al. 2016. Long noncoding RNA Saf and splicing factor 45 increase soluble Fas and resistance to apoptosis. *Oncotarget*. 7(12):13810-26.

their grandchildren added to the infected community. Another generation made it to 1960.

Chloroquine

(vi) However, during the period 1955-60, chloroquine was introduced into the Congo Basin during the World Health Organization effort to eradicate malaria. Chloroquine has the effect of selecting against HIV by suppressing the pro-virus transcription.[22] This naturally selected for viruses (i.e., HIV-1M subtypes) that replicate rapidly.

---

[22] Naarding MA et al. 2007. Effect of chloroquine on reducing HIV-1 replication in vitro and the DC-SIGN mediated transfer of virus to CD4+ T-lymphocytes. *Retrovirology*. 4:6.

Murray SM et al. 2010. Reduction of immune activation with chloroquine therapy during chronic HIV infection. *J Virol.* 84(22):12082-6.

Akpovwa H. 2016. Chloroquine could be used for the treatment of filoviral infections and other viral infections that emerge or

(vii) In the milieu of tropical diseases, no one much noticed HIV-AIDS in the Congo in the 1960s and 1970s, but when the virus was brought to the United States where chloroquine was not a routine daily drug, the HIV-AIDS epidemic became apparent in 1980.

My initial view (2006) had been that chloroquine (which clearly affected the progress of HIV infections) was the drug that had driven the evolution of HIV from the start.

I wrote: [23]

> *'The evolution of the Simian virus (SIV) into a human virus (HIV) is regarded as an artifact. In contrast, a fundamentally different hypothesis has been*

---

emerged from viruses requiring an acidic pH for infectivity. *Cell Biochem Funct.* 34(4):191-6.

[23] Parris GE. 2006. The timing is right: Evolution of AIDS-causing HIV strains is consistent with history of chloroquine use. *Med Hypotheses.* 67(5):1258-9.

Parris GE. 2007. AIDS: caused by development of resistance to drugs in a non-target intracellular parasite. Med Hypotheses. 68(1):151-7.

proposed [Parris GE. Med Hypotheses 2004;62(3):354-7] in which it is presumed that in hyper-endemic areas of malaria (central Africa), all primates (humans and non-human primates) have shared a retrovirus that augments their T-cell response to the malaria parasite. The virus can be called "primate T-cell retrovirus" (PTRV). Over thousands of years the virus has crossed species lines many times (with little effect) and typically adapts to the host quickly. In this model, AIDS is seen to be the result of the development of resistance of the virus (PTRV) to continuous exposure to pro-apoptotic (schizonticidal) aminoquinoline drugs used to prevent malaria. The hypothesis was originally proposed based on biochemical activities of the aminoquinolines (e.g., pamaquine (plasmoquine(TM)), primaquine and chloroquine), but recent publications demonstrated that some of these drugs definitely adversely affect HIV and other viruses and logically would cause them to evolve resistance. Review of the timeline that has been created for the evolution of HIV in humans is also shown to be qualitatively and quantitatively consistent with this hypothesis (and not with either version of the conventional hypothesis). SARS and Ebola also fit this pattern."

And soon thereafter I managed to get the entire hypothesis published (here is the abstract):

Med Hypotheses. ***2007;69(4):838-48. Epub 2007 Mar 26.***

Mechanism and history of evolution of symbiotic HIV strains into lethal pandemic strains: the key event may have been a 1927 trial of pamaquine in Leopoldville (Kinshasa), Congo.

Parris GE

In previous papers, I have rejected both the zoonosis and the serial transfer hypotheses of the origin and evolution of the current lethal pandemic strains of HIV. The hypothesis that fits the critical observations is that all the human and nonhuman primate species in central Africa (an area of hyper-endemic malaria) have shared (through inter-species transfers) a "primate T-cell retrovirus" (PTRV), which has adapted to each host species. This retrovirus is believed to assist primate T-cells attack the liver stage of the malaria infection. Each geographic region has a dominant primate host and a characteristic virus. Starting in 1955 and continuing into the late 1970s, chloroquine was provided by the WHO and used for prophylaxis against malaria. Chloroquine has a number of biochemical activities but two of the most important are blocking transcription of cellular genes and proviruses activated by NF-kappaB and blocking the glycosylation of surface proteins on viruses and cells. Concurrent with the development of resistance of the malaria parasite to chloroquine, HIV strains were quickly selected, which have enhanced transcription rates (by inclusion of multiple kappaB binding sites in their long terminal repeats by recombination) and enhanced infectivity (fusogenicity) (most likely by mutations in multiple viral genes that regulate glycosylation of Env). There also may have been mutations that enhanced activation of NF-kappaB in the host cell. These changes in the retrovirus genome were not manifest in effects of the HIV strains as long as the hosts were under the influence of chloroquine. But, when the virus infects people who are not protected by chloroquine, the virus multiplies more rapidly and is more communicable. Fortunately, most of these strains (i.e., HIV-2 groups, and HIV-1 O and HIV-1 N) self-regulate (i.e., infected cells kill infected cells) well enough that viral loads remain subdued and bystander cells of the immune system are not excessively attrited. In the case of HIV-1 group M, however, there is more going on. Following the work of Korber et al. on the phylogenetics of HIV-1 groups M, I reach the conclusion that the major subgroups giving rise to the worldwide pandemic, were founded in a 1927 clinical trial of pamaquine (plasmoquine) in Leopoldville (Kinshasa). This drug is much more toxic that chloroquine and appears to have strongly selected for resistance to apoptosis in infected cells, which allows these subgroups to attrite bystander cells leading to AIDS.

My Perspective

My perspective is, of course, biased. The hypothesis I articulated between 2004 and 2007 is essentially a conventional description of a fast-evolving parasite developing resistance to a drug. Of course, it is well known that HIV rapidly develops resistance to drugs and that is why a cocktail of drugs has been used to control AIDS since 1997. This is a common problem in therapeutics and it is hard to understand why no one (except me) seems to view the ***origin HIV-1M*** that way. The two academic communities (i.e., *medical treatment* and *viral evolution*) do not seem to overlap very much. Indeed, I suspect that the number of people interested in where the virus came from is dwindling as it has become less lethal.

I also perceive a strong bias against assigning any personal or institutional blame for HIV-1M. The ruthless pursuit by Edward Hooper of the personnel originally involved in the polio vaccine development and testing in Africa, caused a chilling effect on academic investigation. Hooper, being a journalist not a scientist, of course, wanted to defend his hypothesis and has taken every opportunity to criticize and challenge in the public arena scientific findings that unequivocally rule out the plausibility of his hypothesis.

The dirty little secret that western scientific research has frequently chosen Africa or Africans as biological test subjects for bioactive products (e.g., drugs, pesticides) that will be used world-wide is becoming more and more embarrassing. For example, there is a very good film (now video) produced by the US Armed Forces in the 1945 (War Department Film Bulletin No. 195) regarding development and testing of DDT in the US. All the test subjects who are exposed to various insect pest and DDT are black American men. Fortunately, DDT is virtually harmless to humans and we subsequently see white soldiers and children running around in clouds of the pesticide. But I imagine what my students think when I show the video in class. Granted that the Congo Basin has a full spectrum of diseases and provides a rational location for clinical trials and potentially the local people will benefit.

The "cut hunter" hypothesis has the great academic appeal that no one (certainly no one in the medical profession) is to blame. But for reasons stated above, neither it nor the serial passage hypothesis fit the facts of HIV-1M. On the one hand, this is valuable to know because zoonoses are (as shown for HIV) common. And while there is plenty of evidence that existing diseases are spread by introduction of skin-puncture medical techniques into a population where sanitary discipline is not well established and economics weigh against one-use instruments, at least these are known diseases.

The idea that we (collectively) created a very serious lethal disease where none had previously existed is not an appealing thought. Given the well-known problem with iatrogenic, nosocomial infections, it would not be surprising if the clinic in Leopoldville (Kinshasa) was the origin of HIV-1M.[24] But as noted in the case of the serial transfer hypothesis, medical practice was spreading certain viral infections through work in the field independent of any hospital.[25] Indeed, the broad screening programs by Jamot for sleeping sickness in the

---

[24] Hogan CA et al. 2916. Epidemic History and Iatrogenic Transmission of Blood-borne Viruses in Mid-20th Century Kinshasa. *J Infect Dis.* 214(3):353-60.

Pépin J and Labbé AC. 2008. Noble goals, unforeseen consequences: control of tropical diseases in colonial Central Africa and the iatrogenic transmission of blood-borne viruses. *Trop Med Int Health.* 13(6):744-53.

[25] Mbopi-Keou FX et al. 2014. The legacies of Eugène Jamot and La Jamotique. *PLoS Negl Trop Dis.* 8(4):e2635.

1920s and 1930s virtually guarantee that HIV-1M **was not** in an adult population at that time and guarantee that if serial transfer had caused HIV it would have appeared in many places, not just Kinshasa.

The point is that even if this hypothesis is wrong, the failure of mainstream medicine and academia to consider the effects of drugs (especially anti-malarial drugs) in the evolution and spread of HIV-1M is a great mistake, because there are many other diseases that could follow such a path (Ebola and SARs come to mind).

A Final Note

Feel free to distribute this document *with attribution* to me.

If anyone can prove this hypothesis wrong, I will readily modify or abandon it. That's the way science works.

I believe that my use of graphs and quotes from copywritten journals is in the public interest and within the boundaries of Fair Use. If you want to repeat these experiments you must see the journals.

www.ingramcontent.com/pod-product-compliance
Lightning Source LLC
Chambersburg PA
CBHW030737180526
45157CB00008BA/3203
* 9 7 8 1 0 7 9 9 8 9 3 1 1 *